WEEKEND WITH WILLOWS

Willows as a younger man with one of his early gas balloons

WEEKEND
WITH
WILLOWS

LONDON TO OXFORD IN AN OLD-FASHIONED GAS BALLOON

by

SIR ALISTER HARDY, F.R.S.

Illustrated with sketches by

W. ROBERT HODSON

Including
an epic poem
THE BALLOONATICS

A light-hearted version of the
same story written in comic verse by

NEIL MACKINTOSH

ALAN SUTTON
1986

ALAN SUTTON PUBLISHING
BRUNSWICK ROAD · GLOUCESTER

First Published 1986

British Library Cataloguing in Publication Data

Hardy, *Sir* Alister
Weekend with Willows: London to Oxford
by balloon with E.T. Willows.
1. Hot air balloons—England—History—
20th century
I. Title
629.13'092'4 TL638

ISBN 0-86299-241-9

Jacket painting by W. Robert Hodson

Typesetting and origination by
Alan Sutton Publishing
Printed in Great Britain

To the memory of
ERNEST THOMPSON WILLOWS
1886–1926

an enthusiastic balloonist and
early pioneer of British airships who
met his death in a tragic balloon accident

CONTENTS

AUTHOR'S FOREWORD

I have related this story to friends on many occasions over the last sixty years, for it happened all that time ago. Some have expressed amazement that my memory should be so exact, but this isn't really surprising because few events have given me greater enjoyment or excitement; I know it all by heart.

Several friends have suggested that it would make a good chapter in an autobiography. There are two reasons, however, why I have felt that it would be better as a slender volume on its own. For one thing there are many people who are keenly interested in ballooning since the new vogue of the hot-air balloon has taken hold and made the old-fashioned use of gas (coal gas or hydrogen) as a lifting source almost obsolete; they may like to compare the older methods with the new but would not want to read a large book recounting so many other episodes in my life. I should hasten to add, in case I should be accused of lamentable ignorance, that of course the use of

hot air is the older of the two methods. The story is too well known to be repeated, except to say, in case this part is not generally known, how it was that the two brothers Etienne and Joseph Montgolfier, paper manufacturers near Lyons, invented the first balloon. They were sitting by the fire watching the smoke curl up the chimney and at once thought that if they were to fill a huge paper bag with smoke it might carry people aloft; and this is just what they did. When they first went up they used a huge fire of damp straw to make as much smoke as possible, not realizing that it was the hot air, and not the smoke, that carried them aloft.

The modern hot-air balloon is a much more sophisticated machine, the air being heated by turning on a tap to allow a great, intensely hot flame from a cylinder of butane gas to warm up the air in the envelope as soon as its cooling causes the balloon to sink. The violent roar of the butane burner every so often in the course of flight is in great contrast to the old gas balloon, which was absolutely silent; that was one of its great charms. It has largely gone out because nowadays it is more dangerous; the live electric cables which criss-cross the country make landing more hazardous, and flight is less controlled

than with the butane burner; also, there is the danger of fire if struck by lightning, and the inflation of the balloon with gas is much more time consuming.

Another reason for making the book a separate little volume is to dedicate it to that remarkable man Ernest Willows, around whom the story centres; he was not only an enthusiastic balloonist but a most original designer of early airships. Whilst I knew something of his airship work as a schoolboy, as I shall relate, I am greatly indebted, regarding its details, to Mr Alec McKinty's book *The Father of British Airships: a Biography of E.T. Willows* (London 1972); in this he gives a full account of the development and performance of those truly pioneer machines. My book may help in a small way to keep his memory alive by adding to McKinty's account a more personal experience of his enthusiasm. I am also grateful to Mr McKinty for supplying me with the excellent photographs of Willows: one as a young man in his early ballooning days, which is reproduced as a frontispiece, and another on page 20 of him in one of his airships; then in addition he has let me use the photograph of Willows landing his airship in front of the City Hall of Cardiff (page 21).

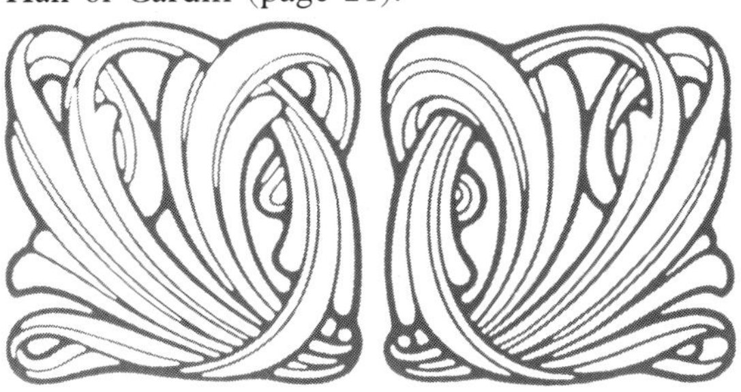

I must now say how lucky I have been to get Mr W. Robert Hodson, to whom I had told the story, to illustrate it with his lively sketches which add so much to my account of the adventure. Then I must thank the relatives of the late Dr Neil Mackintosh, who was one of our aerial party, for allowing me to print, as an appendix, his most amusing version of the story, told in comic verse, in his epic poem 'The Balloonatics'.

Finally I must acknowledge with gratitude the great help I have received from my secretary Miss Anita Jo Dunn in getting the material ready for publication.

CHAPTER ONE

My Meeting with Willows

CHAPTER ONE

This little adventure began on a glorious afternoon in mid-September 1924 at the great British Empire Exhibition at Wembley. It started with an extraordinary meeting which would have had no significance for me had it not been for a remarkable obsession that I developed at my prep school (Bramcote, at Scarborough); so really I should begin with that.

I had two passions as a boy: firstly for natural history, particularly insects, which I began to collect at a very early age; then when I went to my prep school I developed an entirely different interest – almost an obsession – for airships, balloons and flying machines, as we called them in those early days. When I started at prep school, at about the age of ten, my father having died when I was quite young, I found that collecting insects and other natural history objects was almost looked down upon. Boys were mainly talking about motorcars, which were just coming into use by the more wealthy parents,

and also railways, and for a period my natural history studies lapsed until I went on to my public school (Oundle), where the passion for insect study reasserted itself.

I have retained an extraordinary memory for the period of my craze for airships and aeroplanes. One day, being kept in bed with a temperature, I was lent a copy of *Pearson's* magazine – those were the days when everyone read monthly magazines: the *Strand*, the *Windsor*, and others – and in it was an article on airships, which were then called dirigibles, or navigable balloons; and about the same time the first army airship, the *Nulli Secundus*, made a celebrated voyage over London, encircling St Paul's. I became absolutely fascinated. Other boys collected stamps; I started to collect every picture I could find of early airships and balloons. Then, suddenly, there was the arrival of the aeroplane: Wilbur Wright had just arrived with his machine from America, and Blériot and Latham were competing to be the first to cross the English Channel in monoplanes. I still have four large scrap albums of cuttings and photographs covering all the important flights of the years 1908–1910.

Many of the boys were sent copies of the

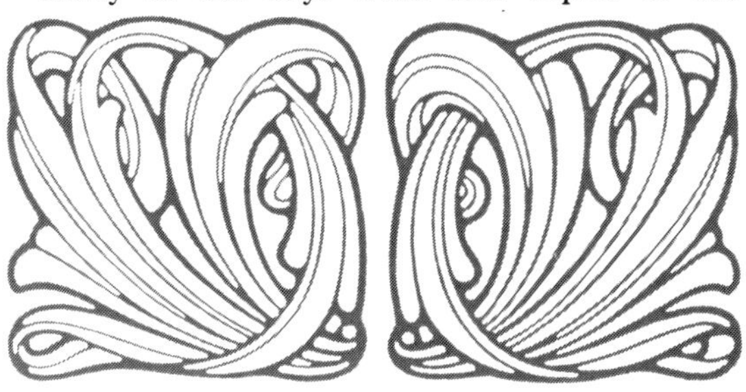

illustrated papers – the *Sphere*, the *Illustrated London News*, the *Graphic*, and so on. Most of them gave me the photographs of airships and aeroplanes when they had finished with them, but others, more greedy, used to do a deal with me, swopping the pictures for my meagre supply of Harrogate toffee or Velmar Suchard chocolate. I also read as many books as I could lay my hands on describing the early ballooning adventures, and I was particularly struck by that daring Brazilian Alberto Santos-Dumont, who, really before my time, excited the whole of Europe with his adventures in quite small airships which he flew over Paris, eventually, after a number of mishaps, winning the prize for the first aircraft to make a complete circuit of the Eiffel Tower.

Now in England, just at the time when I was beginning my collection, a remarkable man, E.T. Willows, began experimenting with very small airships, as Santos-Dumont had, but they were better than his, and Willows' adventures made me one of his most ardent admirers. At the beginning of 1909 there was published the first weekly magazine devoted, as it said, to the interests, practice and progress of aerial locomotion and transport; it was called *Flight*. I at once

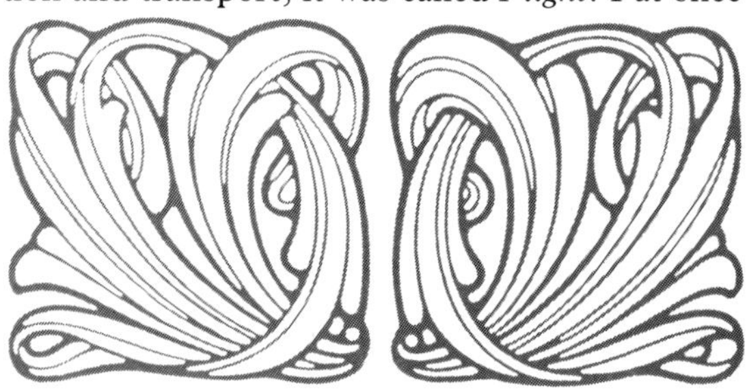

subscribed to it. It was produced and wonderfully illustrated for the modest price of one penny, and each week I looked forward to its appearance. It was there that I learnt, towards the end of that first volume, of the beginning of Willows' experiments.

All this is really to explain why the meeting I had on that September afternoon in 1924 was so extraordinary. Suspended above Wembley was a captive balloon, a kind of kite-balloon, or like a barrage balloon, advertising the *Daily Graphic*. For half-a-crown you could be taken up to, I suppose, five or six hundred feet to get a view of the whole exhibition. I of course joined the small queue of people waiting to make their first ascent in a balloon. It was certainly a wonderful view we got of all the many brilliantly decorated pavilions representing the different countries that made up the British Empire of those days. Then I suddenly noticed a little card stuck to the edge of the basket; it just said: 'Your pilot is Mr E.T. Willows'. I gasped: 'Gracious, it can't be!' and turned to this Mr Willows. 'Surely you must be the great airship Willows, aren't you?' He replied: 'Well, I know of no other.' 'Good heavens,' I said, 'you were one of my boyhood heroes.' He then told me the sad history of how

Daily Graphic balloon at Wembley

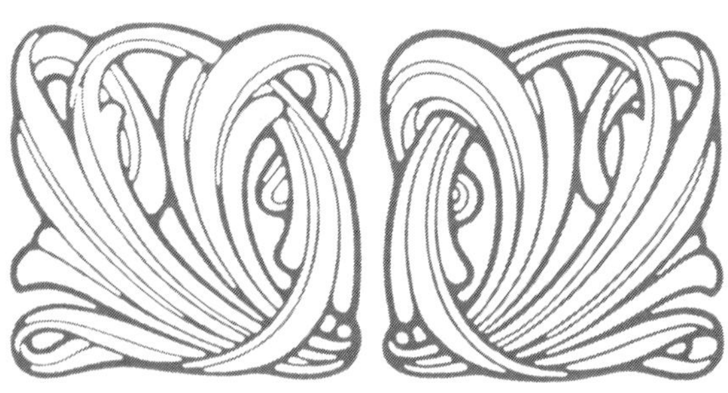

he had spent all his money, and indeed his father's money also – for his father backed him to the hilt – in experimenting with his early airships. They had both become bankrupt and he was now making his living by piloting captive balloons for advertising, or taking people up at various fêtes and shows. It was a wonderful opportunity for me to discuss his early airship designs, about which I had learnt many years before at school from the copies of *Flight*. How excited he was to find that someone really knew about the development of his inventions, because he was indeed the first to patent and successfully use a system of propellors on either side of his little car with which to steer the airship up or down or to right or left. He could manoeuvre his little craft better than anyone else had ever done. The full details of his system are well told in the book by Alec McKinty to which I have already referred. You can imagine that I spent the rest of the afternoon paying out half-crowns for continual trips up and down with Willows whilst we discussed both his machines and his remarkable adventures. To show how manageable his airship was he once landed it without a hitch right in front of the City Hall of Cardiff, his native city. He made long overland

Willows in one of his airships, on the way to Paris.

Willows landing his airship right in front of the City Hall at Cardiff.

trips and was the first man to cross the English Channel in an airship from England to France, eventually making his way to Paris. It was a sad disappointment to him that the government took little notice of his pioneering work, and one cannot help feeling that the little army airships which followed the *Nulli Secundus I* and *II* owed a lot to Willows' designs.

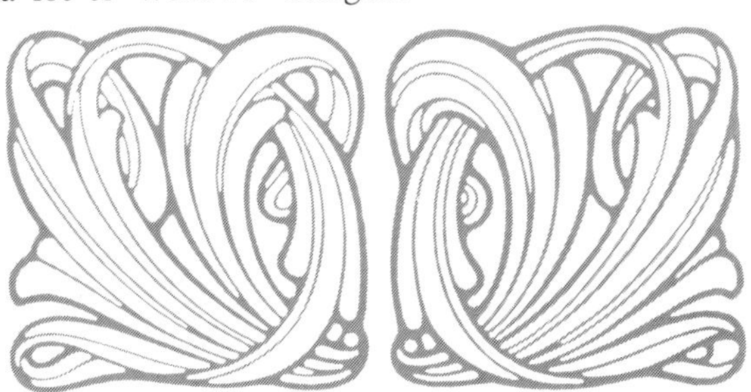

On my last ascent on that afternoon I said to him: 'I suppose it is no longer possible to make a voyage in an old-fashioned gas balloon?' He replied: 'Of course it is! I would love to take you.' I then said: 'How wonderful it would be to cross London by balloon.' 'Indeed you shall,' he responded. 'If you can get three friends to come with you and you each put ten pounds towards the cost of gas, filling of sandbags, etc., nothing would give me greater pleasure than to make such a trip. I am longing to do it. I am sick of this terrible routine captive balloon or advertising work.' I could hardly believe it. I thought the cost would be at least a couple of hundred pounds. It was only long afterwards, in fact after reading McKinty's book, that I learnt how poor he was, and I feel ashamed that we didn't volunteer to give him so much more. He was very keen to make the trip and said that Spencers, for whom he worked, would lend him one of their best and biggest balloons. So it was that our little adventure was set in motion.

I was then in London, preparing for an expedition to the Antarctic in Scott's old ship *The Discovery*. It was to be the first of a series of voyages south to study everything in the South Polar seas relating to the great whales there

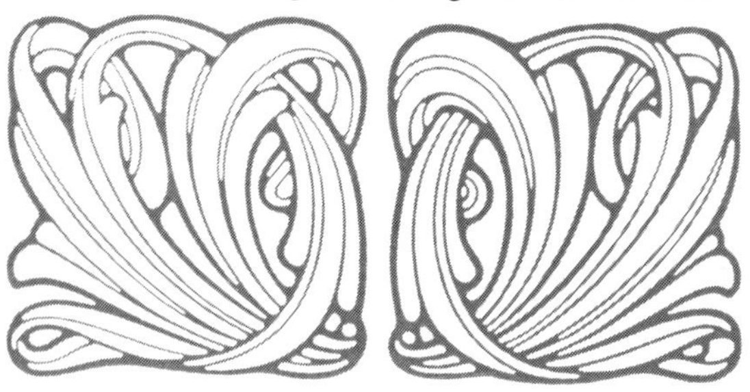

being fished. I got Neil Mackintosh and John Wheeler of the expedition to join me, and the third was one of my very oldest friends, Geoffrey (later Sir Geoffrey) Vickers, who was awarded the V.C. for gallantry in the First World War. Willows arranged to visit me in my digs in Chelsea to discuss the plans. As I have said, I very much wanted to cross London. 'What about the wind?' I asked Willows. 'That is quite simple,' he said. 'In the old days, when I made many trips, I had an arrangement with all the gasworks around London to supply me with some 80,000 cubic feet of gas if the wind should be coming from their direction. I will make a similar arrangement.'

When he came to see me I showed him my great collection of airship, balloon and aeroplane pictures and he was delighted to see several photographs of his early airships, which he kindly autographed for me.

The trip was planned for a Satuday afternoon in early October, in fact on October 4th. We arranged to go up at about half-past twelve.

Autographed photographs of Willows' early airships

CHAPTER TWO

Planning the Voyage

CHAPTER TWO

When the morning came it was absolutely still; all the smoke from the chimneys was rising vertically into the air. I rang up Willows and said: 'What can we do? We cannot cross London with no wind at all.' He said: 'That's all right. In case this might happen I have made arrangements with the gasworks next to the Oval Cricket Ground in the centre of London, so that at any rate we can cruise slowly over the very centre of the city.'

When we got to the gasworks we expected to see the balloon fully inflated and almost towering above the gasometers, for it was going to be a very big balloon; but there was no sign of it. We found a very angry Willows packing up the deflated balloon. A little breeze had sprung up from the east and the manager of the gasworks refused to let Willows go up because he said he would damage his gasometers, which were immediately to the west of the space where the balloon had begun to be inflated. Willows said

he was certain that by letting out ballast at once he could easily jump over the gasometers, there was no danger at all. However, the manager was adamant. Willows then told us he had made another arrangement with the South Metropolitan Gas Company at their big gasworks at Bugsby Marsh to the east of the Isle of Dogs, and that he was going to take the balloon there, adding that it would probably be five o'clock or so before it was fully inflated.

Willows, always cheerful and enthusiastic, said 'We shall now cross London by night, or at sunset, which will be even more wonderful than crossing it by day.' We had never expected to go up at night. Geoffrey Vickers and John Wheeler were a little concerned because they had not told their wives that they were going up in a balloon and this meant that they would not get back until very late at night. However, I think they made some excuse by telephone and all was well. Willows of course was organising all the arrangements and wouldn't come to lunch with us, but the rest of us went off and had a very good meal at a hotel famous for its whitebait suppers, and then proceeded to Bugsby Marsh. We found the balloon only about a quarter inflated. It looked, as Geoffrey Vickers said, just like a great

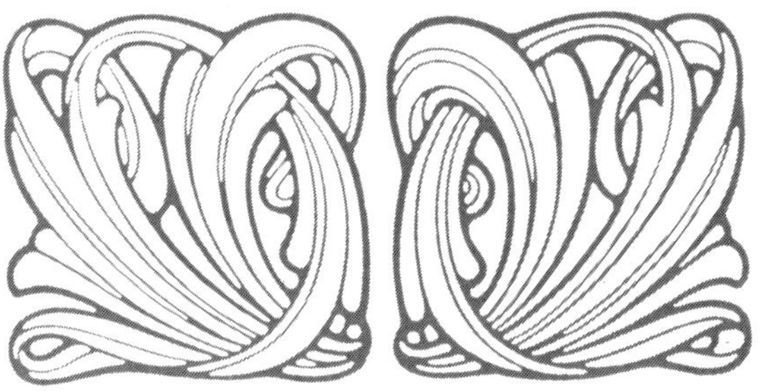

The beginning of the inflation of the balloon.

stranded jellyfish. As it was being filled it grew and grew, and all the time the men helping would keep adjusting the sandbags which were hooked on to the netting covering the balloon all round, continually putting them into the mesh further down until gradually the whole balloon began to take shape. It was indeed after five o'clock before it was fully inflated. It was a wonderful evening, slightly hazy, the whole sky a vivid orange, whilst the sun, like a ball of fire, was slowly sinking in the west.

Before we actually take off I think it will be best if I give a brief description with diagrams of the anatomy of the gas balloon. In olden days

A later stage in the inflation. As the sandbags, which are attached to the netting covering the envelope of the balloon, have been systematically lowered, the balloon fills and tends to rise.

balloons were usually pear-shaped, but latterly they have become more spherical. The balloon is a gigantic sphere made of rubberised or otherwise treated silk or other fabric. At the bottom it has a tubular opening for its inflation and through which two cords pass – one to a spring valve to control the volume of gas, and the other to what we call the ripping-panel. The latter is used when finally landing, as one has no further use for the gas. It acts rather like a zip-fastener, which on being pulled opens a great slit in the balloon, allowing the gas to escape just before the basket is due to touch the ground. This

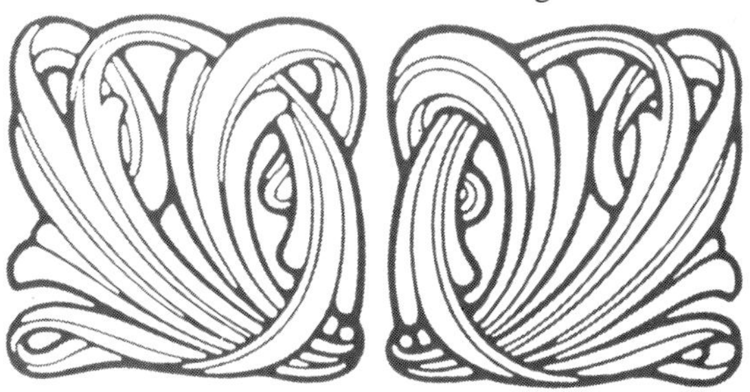

Coming to the last stages in the inflation of the balloon. The sandbags are now attached by long cords to the netting and firmly tied to the ring above the basket.

means that one lands comfortably; the whole balloon is blown gently to one side, emptying as it does so, and it is then ready to be rolled up for packing. The balloon itself is enclosed within a great bag of strong cord netting with wide meshes about a foot across. As the netting reaches below the equator of the balloon it narrows and at its edge each mesh joins a stout cord passing down to be firmly tied to a metal ring which is usually slightly smaller in diameter than the width of the square basket suspended below it. The basket indeed hangs from this metal ring by at least eight strong ropes, and round the basket hang the bags of sand which act as ballast.

The art of ballooning is to balance one's supply of sand with one's amount of gas; if one is rising too fast the valve at the top of the envelope is opened by pulling the cord to release a little gas, and if one is falling, a little sand is thrown out from one of the bags. On the outer edge of the basket there is a longish rope, perhaps about 100 feet, coiled up, and at the end of it is an anchor-like grapnel. If necessary, this is thrown out to anchor the balloon in landing. Then at the bottom of the basket, also coiled up, is a longer rope of some 300 feet; this is known as the

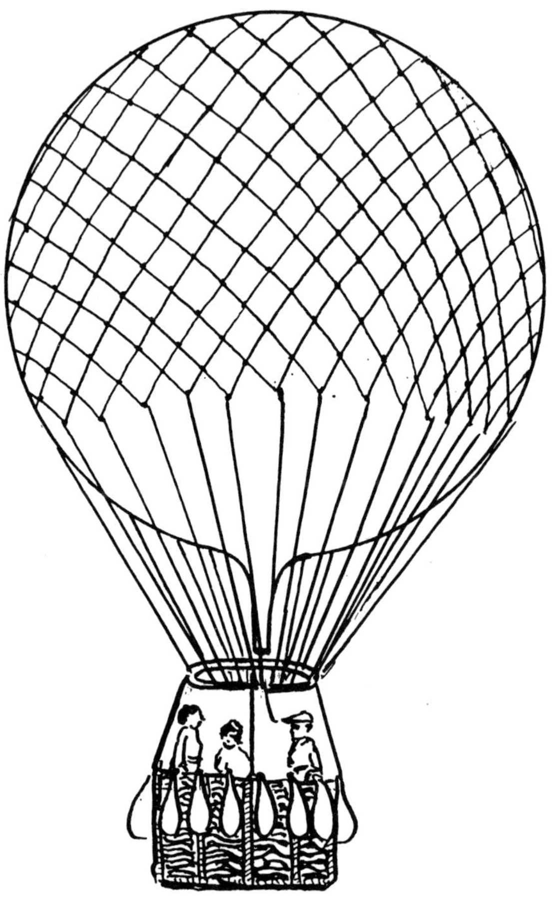

Two diagrams of a typical gas balloon. On the left, seen from the outside; on the right, a section through the balloon showing the cord attached to the valve at the top of the balloon and another cord attached to the ripping panel.

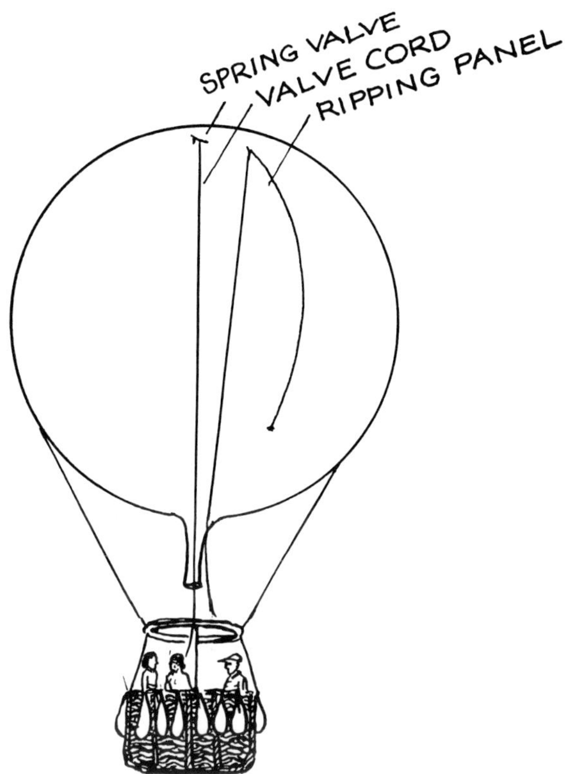
SPRING VALVE
VALVE CORD
RIPPING PANEL

guide-rope and can be used for low flying, keeping the balloon at some 250 feet without using up ballast. As soon as the balloon falls a little, the increased weight of the rope on the ground tends to make the balloon rise again, and if the balloon rises too much then the weight of the rope taken up from the ground will pull the balloon down again; so one can travel quite a long way at a balanced height with this trailing guide-rope.

Alongside my sketch of a typical balloon I have a similar diagram showing a section of it. You will see one cord from the basket going straight up the middle of the balloon to the spring-valve at the top; on pulling this the valve opens like a little trap-door and springs shut again as soon as it is released. The other cord, as I have already explained, passes up to what is called the ripping-panel; if one is landing at the end of a flight and decides to go no futher, then just as one is approaching the ground this cord is pulled and it at once opens a long slit in the side of the balloon to release the gas. If there is the slightest breeze the balloon falls flat, clear of the basket, ready for folding up.

CHAPTER

THREE

The Crossing of London

CHAPTER
THREE

After examining everything Willows was satisfied that all was ready for flight, and he clambered into the great basket, which is really like a very large and strong laundry basket. He had given instructions to those who had been fixing on the sandbags to hold on to the basket until he gave the call 'let go'. He then helped us to climb in and we were now ready. A final check and then he shouted: 'Let go!' We at once rose heavenwards – and there came a great surprise to me. It was only in the first few moments that I had the sensation of really going up, as though in a lift. Almost at once – although I think some of the others did not feel it as I did – it seemed as if we were not actually rising up but that the earth was rapidly falling away from us. I remember the others saying, in surprise, that my first remark was: 'The extraordinary security of it all.' One really felt that the balloon was one's little universe and the earth something quite separate that we had

Looking down at the spectators as we begin our voyage

left behind. The little crowd of people who waved us off stood in a circle as we rose gently into a slight wind, which was blowing almost exactly from the east. Below us was the great

Rising and passing over the London Docks as the sun sets in the west

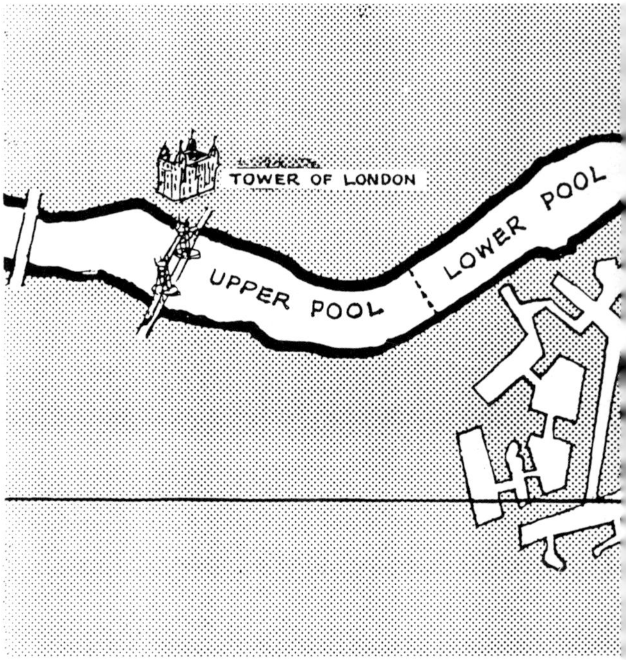

Map of our route across London from east of the Isle of Dogs, on the extreme right . . .

sweep or circle of the Thames surrounding the Isle of Dogs, which in those days was the heart of dockland. The West India Docks were full of shipping, and, particularly beautiful, in one of

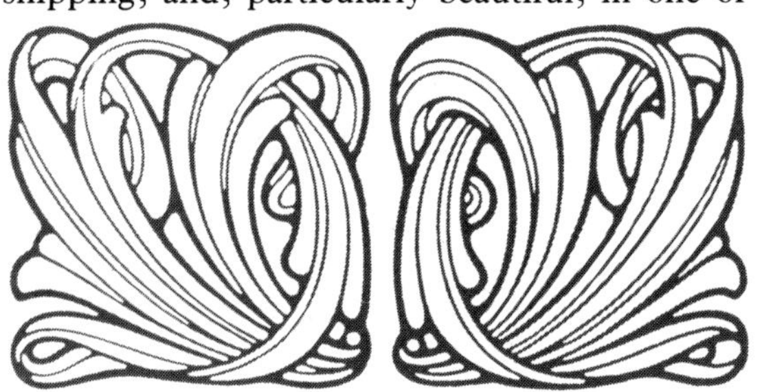

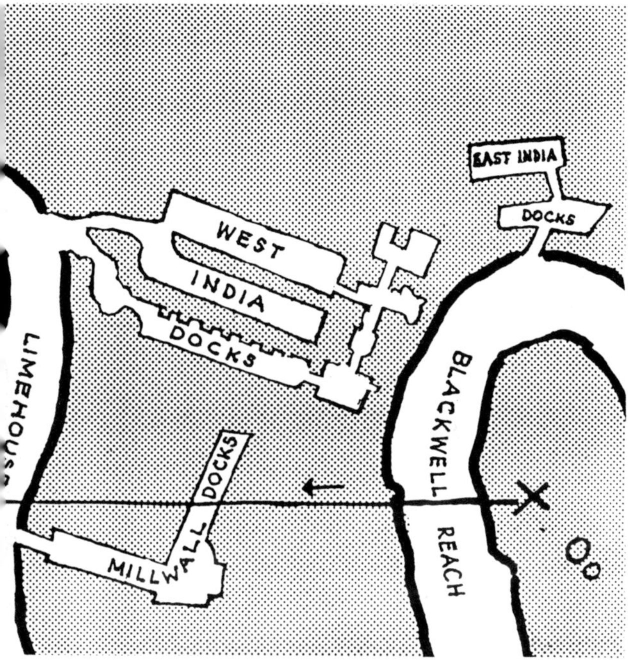

EAST INDIA
DOCKS
WEST
INDIA
DOCKS
BLACKWELL
REACH
MILLWALL
DOCKS

them was a square-rigged sailing ship, with its sails furled, of course, but the yards, masts and rigging standing out in silhouette against an absolutely golden background. The docks were

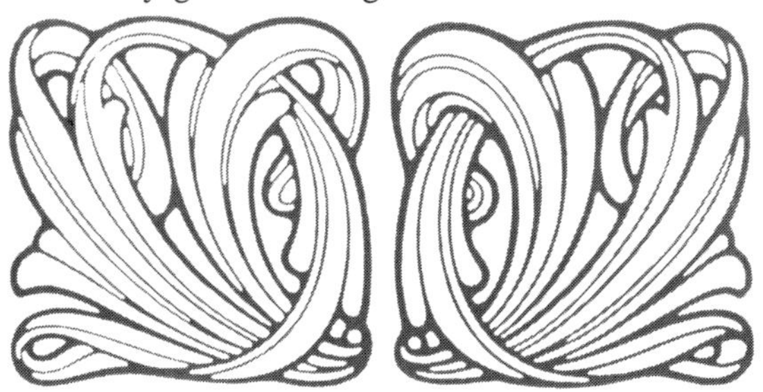

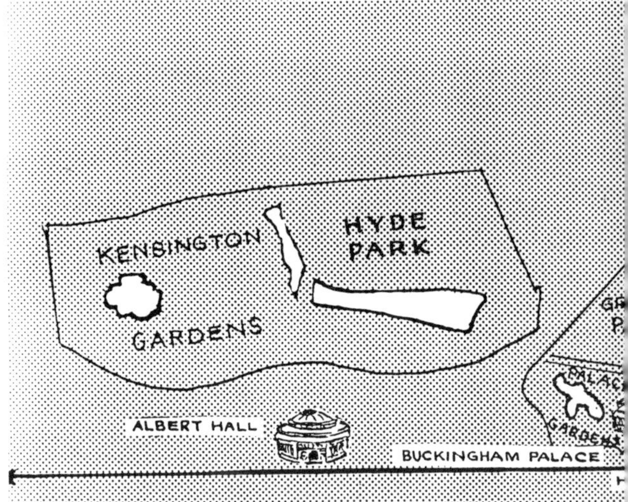

. . . to the Albert Hall and beyond, on the extreme left

reflecting the orange sky which, as I have mentioned, covered the whole scene, with the sun just setting away to the west.

If you take a map of London and put a ruler on it from the South Metropolitan Gasworks, which is marked – or used to be on many of the

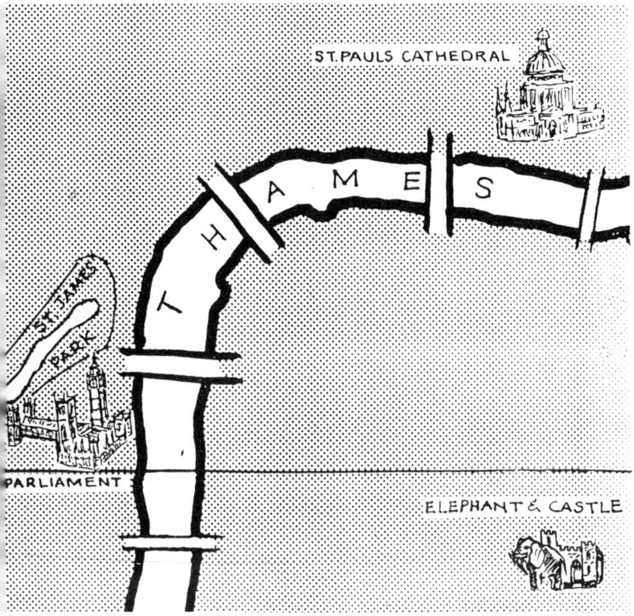

ST. PAULS CATHEDRAL
THAMES
ST. JAMES PARK
PARLIAMENT
ELEPHANT & CASTLE

maps as Bugsby Marsh – and draw a straight line to the Albert Hall you will trace what was almost exactly our course of flight. I remember Willows being very pleased that he could use the gas from the South Metropolitan company because he said it had more hydrogen in it and

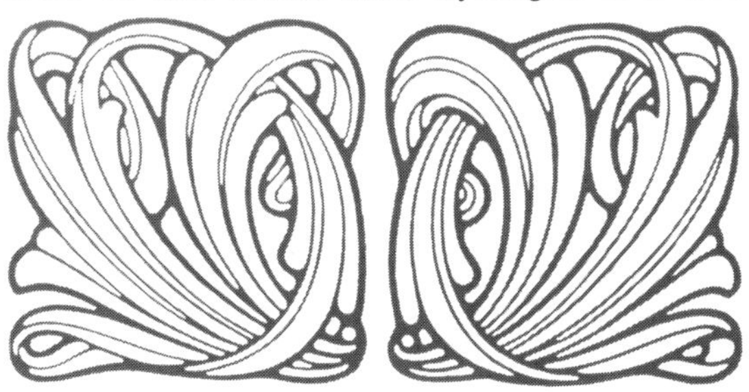

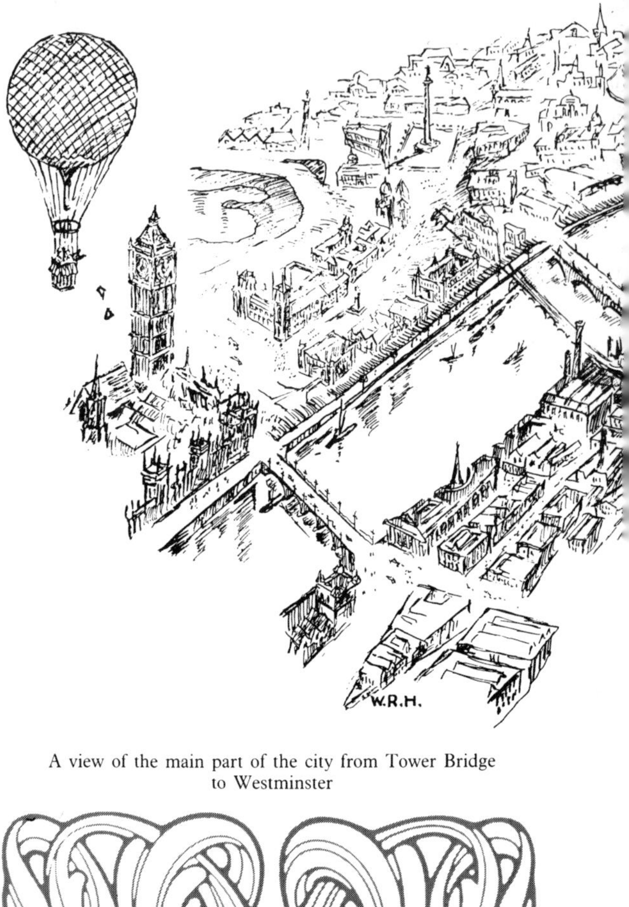

A view of the main part of the city from Tower Bridge to Westminster

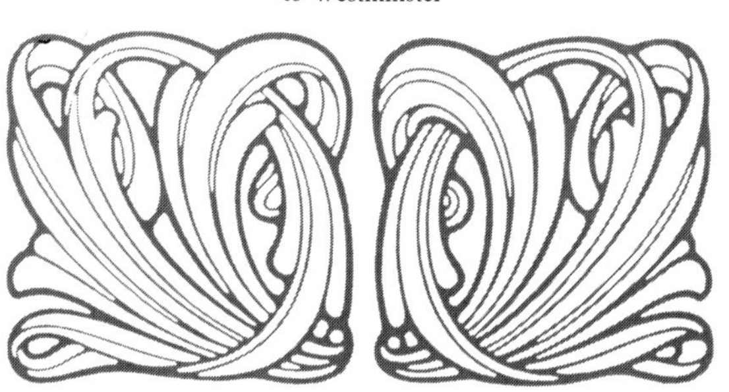

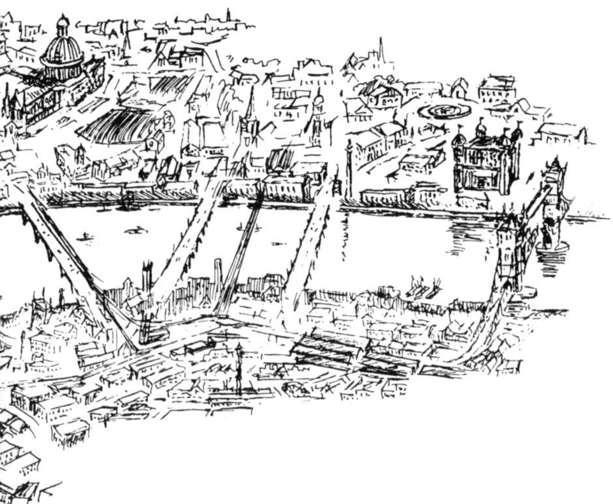

therefore gave greater lifting power than gas supplied by any of the other London companies.

As we drifted slowly over London I was surprised to see that all the time he was talking Willows was letting out cigarette papers.

This was his method of navigation; if the cigarette paper tended to fall slightly he gave a little jerk on the valve, and if they went up he threw out less than a handful of sand (see drawing on p.44). In this way he balanced the

As we float to the top of Big Ben it strikes six

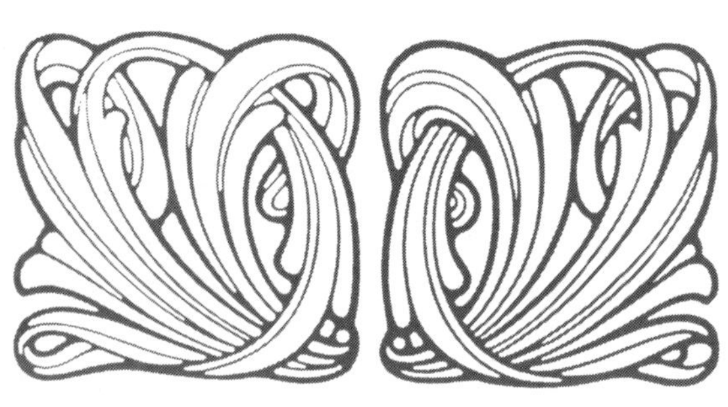

balloon at about 300 feet, giving us a most wonderful view of all the streets below us. To the north we could see Tower Bridge and the Tower itself, and a little further on St. Paul's; but immediately below was the 'Elephant and Castle'.

It was Saturday night and there were great crowds waiting for opening time; everyone cheered as we passed over and we waved as we went along. We seemed to be just above the housetops.

If you follow the line that I have mentioned you will see that we passed just south of the Houses of Parliament, and as we went by Big Ben struck six. In the meantime street after street was suddenly lighting up: Whitehall, Piccadilly, Regent Street, Oxford Street, and so on. The bridges over the Thames were also lit up. Continuing the line of flight you will see that we almost passed over the top of Buckingham Palace, and so on westward until we came down very low over the top of the Albert Hall, so low that we had to throw out more than a little sand and could hear it dropping on to the roof; and there, beyond it, was the great expanse of Hyde Park and Kensington Gardens.

'Well', said Willows, 'you have had a good

look at London close down, now we might go up a bit and get a wider view.' Dropping out a little sand we rose slowly until the whole of London opened out – a maze of lighted houses and streets in all directions. Up and up we went, and now I was delighted to experience what so many of the old balloonists said: the last sounds you hear as you get higher and higher are the barking of dogs and the whistles of railway engines. It was perfectly true; it was indeed surprising how many dogs you could hear barking and how many train whistles there seemed to be, but of course we heard them from over a considerable distance. Today, I suppose, you wouldn't hear the diesel engines with their quite different and less shrill note.

CHAPTER FOUR

En route for Oxford

CHAPTER
FOUR

On going higher we then passed through a thin layer of haze or light cloud, only a very shallow layer it was. Coming above it we found ourselves riding over what seemed to be a silver sea. There was half a moon shining and the whole of this cloud floor was brilliantly illuminated with silver light. To my astonishment I suddenly saw what seemed to be a magic island floating level with us, splendidly lit up. I said to Willows: 'Good heavens, what is that?' 'That is Wembley,' he replied. There were gaps in the cloud, and as we ascended the horizon tended to rise with us, and here was Wembley, some ten miles to the north, showing through a rift in the layer of vapour and appearing to be level with us as an island on this silver sea.

After a time Willows said: 'We had better get down below the cloud now and see where we are going.' So we valved a little, came through the thin layer and saw the whole countryside below us; woods appeared as dark shapes and roads

Floating over a moonlit layer of cloud as if over a silver sea

were like pale ribbons running in different directions, with here and there little groups of lights where there were villages. I got out the map I had taken with me and looking at it tried to imagine what villages they might be, but the trouble was that after looking at a group of lights, then at the map, and then back again to the lights, the village was no longer where I had thought it to be. The balloon, in fact, was very slowly rotating. This is a very common phenomenon. Willows saw what I was trying to do and said: 'I never use a map, but always ask the way.' So we went on a bit further and coming

down quite low saw a farmhouse with all its lights on and some people in the yard or garden

Looking down on the countryside at night

in the front of it. Coming lower still we called: 'Where are we?' There was a scream, and the

Sailing low over the streets of High Wycombe

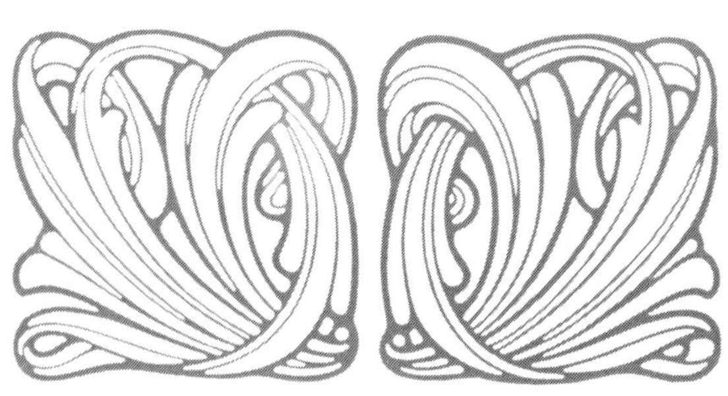

people, who probably could not see the balloon against the layer of cloud, perhaps heard only the voice calling from heaven. They seemed to rush indoors, so we went on, none the wiser.

Now it appeared to me, extending our line westwards, that we must be heading almost exactly for Oxford, and I thought what a wonderful thing it would be to try and land in Tom Quad. This is of course the largest quadrangle in Oxford, that of Christ Church. Presently, ahead of us we saw quite a large town, a mass of lights, and what appeared to be two roads running to the north. I thought, by jove, perhaps they are the Woodstock and Banbury Roads and this *is* Oxford. We were very low indeed and right over the main street, which looked almost like Carfax. Geoffrey Vickers called out over the basket: 'Say guys, is this Europe? Where are we?'

Some sportsman shouted back: 'This is High Wycombe.' 'Good gracious, the Chilterns!' cried Willows, letting out a lot of sand. I had not been to High Wycombe before, but I have since and seen the large number of slender factory chimneys that at any rate were there then. We must have passed close between them. Going out beyond the town we saw a great dark patch

We nearly collide with the church tower

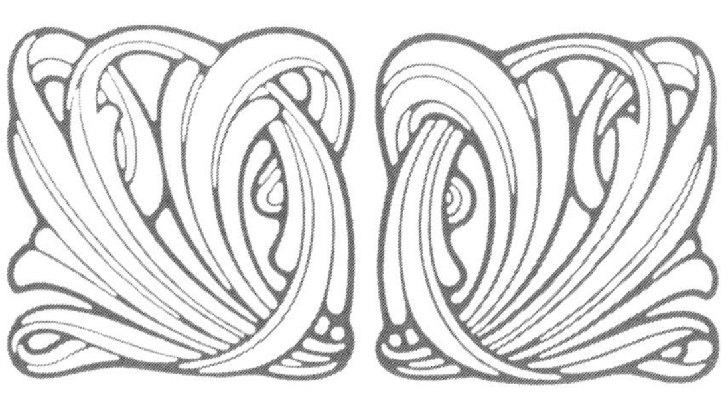

which I thought must be a wood, but then to my astonishment the wood got bigger and bigger and bigger. I pointed it out to Willows, who said: 'Heavens, it is the top of a tree' and called out, 'hang on to the ropes and bend your knees, we are going to hit.' He had told us before we started that it is almost impossible to hurt yourself in a balloon, even if you fall quite fast; you hang on to the ropes, lift your legs up and bend your knees, the basket taking the whole shock of impact, so that you fall as if from a few feet into the bottom of the basket. This indeed was what we did. I thought we had stopped. It all seemed very quiet. I looked over the edge of the basket and to my amazement saw a little string of lights that seemed to be hundreds of feet below us. It was a train. We had obviously bounced on the top of a hill and passed over a deepish valley. Apart from the aneroid barometer the only other instrument we carried was what Willows called his 'bubbleometer'. I never saw him look at it, nor do I know what it was for, but we lost it on that impact. However, it did not seem to worry Willows – the aneroid was the important instrument and it was still securely fastened. It wasn't long before another surprise struck us. Looming through the gloom, at the same height as we

were, was a church tower, and it was coming towards us. We at once threw out more sand, narrowly missed the church and heard the sand falling on the gravestones round about it. I have since visited the church, which stands right on the top of a ridge of the Chiltern hills in a village called West Wycombe.

It was now getting quite late and Geoffrey Vickers and John Wheeler, who hadn't told their wives they were going ballooning, said they must try and get back to London as soon as possible. So we cruised along in the direction of Oxford, and I still hoped we might make it, but the wind was now a very gentle one and we were travelling slowly. We wanted to come down near some village where our two companions might telephone their wives and then get a car into Oxford to catch a train to London; but for a long time we could see only isolated farms. Then a village appeared not far away and, close by, quite a large field which would make a good landing place. By this time there had arisen one of those autumn mists that hang over the Thames valley; it was probably only a few feet in depth, because we could still see hedges and trees, but the rest of the ground was white. We valved and came slowly down, expecting to make a perfect and

We landed with a great splash!

gentle landing. The surprise was that we landed with a great splash! It wasn't mist at all, but the flooded Thames. We had to get rid of nearly all the sand in bags in order to clear the water, which soaked the bottom of the basket and obviously made it much heavier. We drifted on towards a wood, but we couldn't quite clear the upper branches of the trees and crossed it with the basket bouncing from tree top to tree top, birds flying out in all directions.

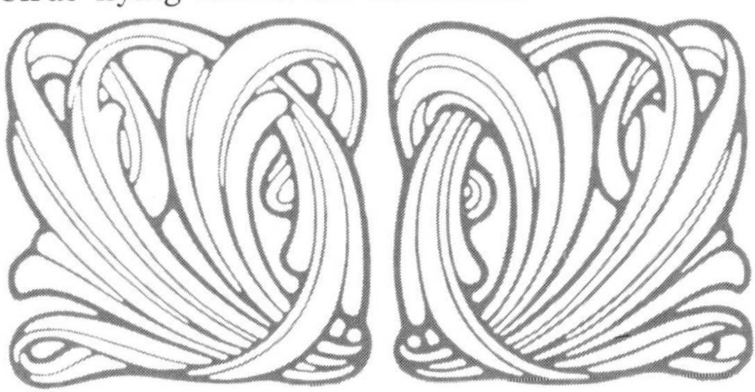

Dragged through the treetops of a wood

Presently we came to another field, a smaller one. Willows said this would do because we could throw out the anchor and bring her down; we did so, but no sooner had we released it than we heard the sound of galloping horses below us. We thought for a moment that we might have caught one on the anchor, but luckily it turned out that the animals were in the next field. We descended very smoothly, the anchor holding fast in the middle of the field. Then Willows astonished us by saying: 'As two of you [Geoffrey Vickers and John Wheeler] are going back to London, we can tie the balloon up for the night and the rest of us [Neil Mackintosh, Willows and myself] can go on tomorrow if we can get some help with re-filling our sandbags with soil.' He then added: 'If one of you would get out of the basket and hang on for dear life to the edge of it we shall be able to move step by step towards a tree and tie up for the night.' Geoffrey Vickers at once got out, and while he hung on to the edge of the basket it was quite easy, little by little, to move the balloon. As soon as he released his weight the balloon went up slightly, and when he put his weight on to it again, down it came. In this way we proceeded towards the edge of the field, in fact towards a

gate, in a series of hops. As we got nearer we saw a large stone roller; this was an excellent thing to tie the balloon to.

Willows said he would stay with the balloon all night, he was accustomed to sleeping in a balloon basket. Fortunately he had some waterproof sheets, otherwise he would have been sleeping in a soaking wet basket. We arranged to go off to the nearest village to enable Geoffrey Vickers and John Wheeler to get a car to Oxford and so to London. Neil Mackintosh and I would stay, if we could, at some local pub. As we came out of the gate two farmworkers on bicycles came along. We stopped them and asked them where we were. 'Waterperry,' they said. We told them we had just come down in a balloon, which in fact could be seen towering above us. We thought they might express some surprise, but not a bit of it, they seemed to accept our arrival as quite normal. On asking them where the nearest town was they replied: 'Wheatley, two miles off.' We then enquired: 'Would you like to earn a little extra pocket money by coming back in the morning and filling up our sandbags with soil?' They readily agreed and said they would be along about ten o'clock. So we set off in the direction they had pointed out. On the way we

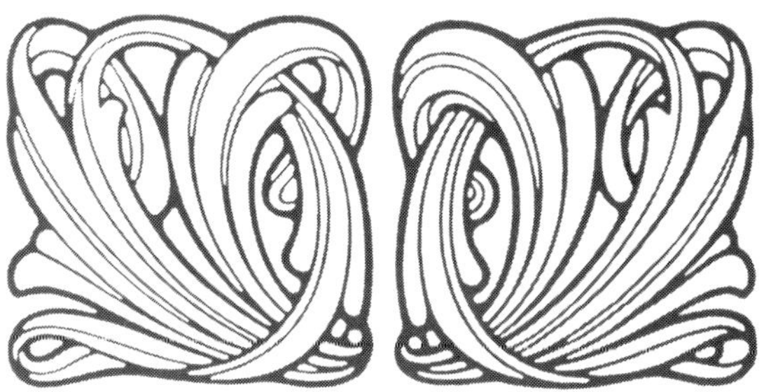

Tied to a stone roller for the night

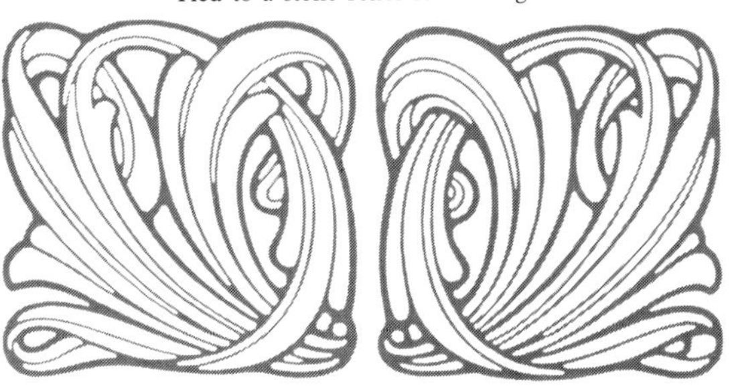

came to a fork in the road and weren't quite certain which way to go. There was a light in a cottage nearby, so we knocked on the door, and an old woman came out. As we stood there we must have looked rather like ruffians because we were wearing all kinds of strange garments. We explained that we had just come down in a balloon. She banged the door in our faces, so we went on, luckily taking the right road, and soon came to Wheatley. Geoffrey Vickers and John Wheeler telephoned their wives, confessed they had been up in a balloon, and would be rather late getting back as they had to get a car to Oxford and then catch a train at about one o'clock in the morning. Neil Mackintosh and I went to the local hostelry. Fortunately we were able to book two rooms and asked to be called on the early side so that we could have breakfast as soon as it could be served. We would then get Willows to come and use one of our rooms for a wash, and so on, and to have his breakfast whilst we looked after the balloon. All worked out as planned. Willows said he had slept quite well. I could hardly believe him, but he was a remarkable man. So off he went to Wheatley for a clean-up and breakfast. Meanwhile the two yokels had turned up and it didn't take them long to fill all our bags with fine, sand-like soil. Also, news of the arrival of a balloon at night had drawn a number of spectators who began to crowd round. By the time Willows returned and checked over everything it was nearly twelve o'clock before we were ready to take off again.

CHAPTER

FIVE

Above the Clouds at 10,000 feet

CHAPTER FIVE

On Willows' return we carefully walked the balloon to the end of the field. Now, to our surprise, the wind, though light, had almost completely changed round and was blowing from the west. Willows wanted to make as slow an ascent as possible, so when we got into the basket we carefully let out soil so that the balloon rose gently, passed down the field and just cleared the hedge. It was a day of sunshine and huge white cumulus clouds sailing in the sky, and we rose steadily.

Neil Mackintosh, for this part of our journey, kept a careful record of our height, shown on the aneroid barometer, and graphed it against time. The diagram he made afterwards, which I reproduce, shows our somewhat erratic journey. We rose gradually to meet the first layer of cloud at about 5,000 feet, and here we experienced the glory of the old-fashioned balloon – absolutely silent as it sailed over beautiful Oxfordshire, the whole countryside below laid out in a checker-

board of fields, woods, larger roads and little lanes. The trees were beginning to show their autumn colours. We then passed up between great masses of cloud, and up again towards another layer above. One realised, for the first time, what a remarkable surface the cloud has. One speaks of a cumulus cloud as being like a great mass of cotton-wool, and indeed its surface seemed just as sharply defined as if it was. As we floated up in the sunlight that came to us through the gaps between the clouds above we could see our shadow most distinctly outlined on the curving surface of the cloud below. We went on steadily rising through this second layer until we came above it into the clearest blue sky, with the sun beating down on us. We continued to climb, and as we did so the clouds began to thicken, and before long our view of the ground below was cut off. It all seemed to take much longer than Neil's chart shows that it did.

It was about 12.40 p.m. when we cleared the upper layer at about 8,000 feet. We had brought sandwiches, pears and bottles of beer for our lunch, and I shall never forget the joyful experience of cruising above that great mass of cloud while we relished our sandwiches and beer. Throughout our lunch Willows regaled us with

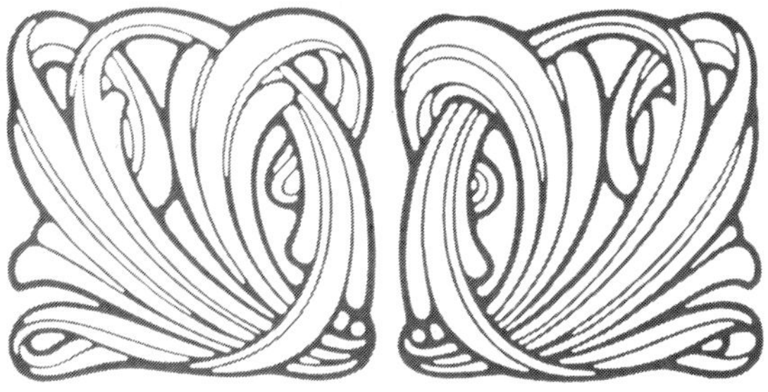

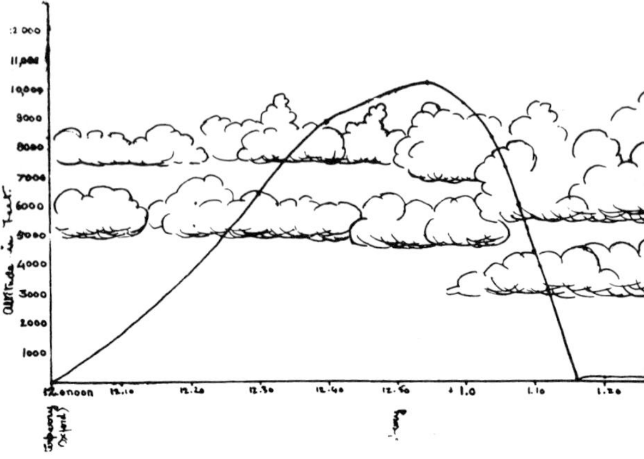

Chart of altitude and time prepared by Neil Mackintosh

stories of his early ballooning, recalling in particular some of his adventures in his little airships. He told us how he had crossed the Channel, eventually to make for Paris, but as he approached the French coast the wind increased considerably. An assistant technician who was with him had all the maps for the trip and was looking at them, trying to make out which point on the French coast they were about to cross, when a sudden gust of wind carried the whole lot

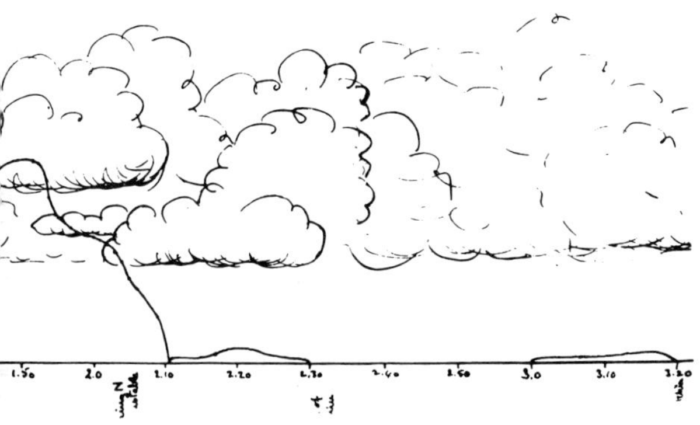

away. There they were, approaching France with all their maps gone.

We went on, up and up, and during this part of our ascent I took the photograph which shows the great sea of cumulus cloud over which we floated, until we reached a height of some 10,500 feet.

Willows then said: 'I don't want to be above clouds for more than a short time. You never know what may be happening; strong winds can

Sailing above the clouds at 10,500 feet

get up and you may be quite unaware of them. You might, for instance, be in a 50 m.p.h. wind and if you hung a handkerchief out it would hang limply, unless you were struck by a sudden gust, as you are in fact part of that wind as you are travelling in the moving air mass.' He pulled the valve-line and we began to descend. I have said how warm it was up there in the bright sunshine; the sun was indeed beating on the balloon, making it like a hothouse and expanding the gas, but as soon as we came into the great mass of cloud the atmosphere at once became exceedingly cold. I began to feel a pain in my ears. I looked at the hand of the aneroid and saw

Falling at 1,000 feet a minute

that we were falling at close on 1,000 feet a minute. We let out ballast as fast as we could, to try and check the fall, but the soil, instead of dropping, shot up past us. Looking up, we saw the underside of the balloon flapping hard in the tremendous updraught as we fell. Because of the loss of gas from the day before – the equivalent of the weight of both Geoffrey Vickers and John

Wheeler – the balloon, instead of being nicely spherical, was more like a great parachute, its underside billowing strongly in the rush of air. We then let out the full 300 feet of the guide-rope, which we thought might help to reduce the speed of our fall as we came to earth. We shot out of the bottom of the cloud to see another layer below us, going in a slightly different direction.

At different heights winds are often travelling in different directions, and therein lies one of the arts of ballooning. In a race, such as the Gordon Bennett international race, the good balloonist will go up and down, carefully using his ballast and gas to find, with the minimum loss of gas, the wind which will carry him in the direction nearest to that which he wishes to take.

Going through this second layer of cloud we were still falling at the same speed. I vividly remember the little village below us, with groups of people and a line of cottages which I had a dreadful feeling we might be going to hit, but luckily we were carried a little way beyond them. I really thought that we should break our legs, but Willows, always cheerful, said: 'You won't hurt yourselves, hang on to the ropes and bend your knees.' The long guide-rope had coiled up

on the ground below us. I remember seeing a group of people come running through a gap between the cottages, evidently expecting that we were about to be killed. As they came hurrying towards us we hit the ground with a tremendous crump and we all dropped to the bottom of the basket. But Willows was perfectly right, we did not hurt ourselves at all. Then came another surprise. Because we had let out so much ballast, and with the whole weight of the guide-rope taken on the ground, the moment we hit, the whole balloon shuddered and at once went shooting up again, much to the surprise of the people who were running towards us. We waved our hands and hats at them as we went up to some 200 feet or so, and here the guide-rope came into use.

I have never discovered what village it was, but it was probably near Luton. We now drifted along with the last ten feet or so of the guide-rope trailing on the ground, balancing the balloon at an even height; across fields and over hedges it went, sometimes being clutched by farmworkers who, thinking that we were intending to land, tried to pull us down. Willows shouted to them to let go. On one occasion they didn't and the balloon dragged the rope through

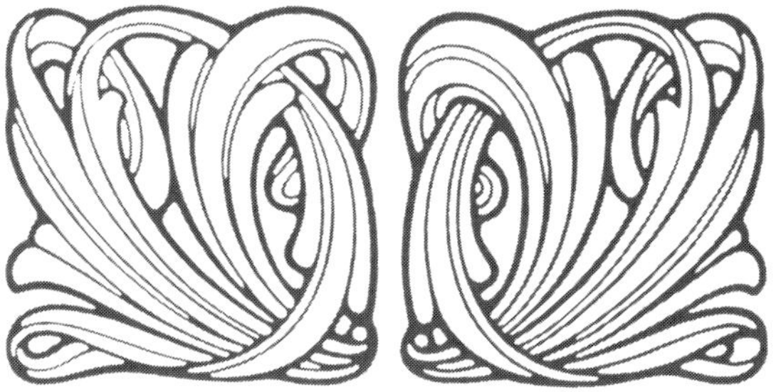

a hedge, where they were scraped off. A little later a farmer caught hold of the rope as it crossed his farmyard. We shouted to him to let go, but he held on. As we approached his farmhouse we had to let out quite a bit of soil, so that the balloon would carry us safely over. We warned him at once to let go or he would be carried up with us. He didn't do so at first, and although I am not absolutely certain, I think he fell into a manure heap.

CHAPTER SIX

Caught in a Thunderstorm

CHAPTER SIX

It was now that we realised the clouds were building up enormously, some looking very dark and threatening. We considered coming down, but Willows didn't like to stop until he was near some town that would have a train to take us back to London. However, two flashes of lightning made us decide to land, but no sooner had we pulled the valve line than somewhat to our dismay we were immediately carried upwards.

It was clear that we had got ourselves into one of the great updraughts that occur on the edge of a thunderstorm. In spite of holding the valve open we were rushing upwards. There were peals of thunder and rain to one side of us, but luckily not on ourselves. Our speed of ascent slackened off at about 5,000 feet and we thought all was going to be well when unexpectedly we were caught in another updraught and carried up to nearly 7,000 feet, with a further outbreak of thunder to the side of us. You will see the

Being drawn up on the edge of a dark thundercloud

photograph that I took as we were rushing up the side of this great black cloud.

It seemed extraordinary that we should still be going up in spite of the valve being open. Then, just as we were really getting somewhat alarmed, the balloon began to drop and we descended almost as quickly and in a matter of a few minutes lost 4,000 feet. After slackening a little we began to drop again and fell as steeply as we had done in the earlier quick descent. According to Neil's chart we hit the ground about ten minutes past two o'clock. Again by following Willows' instructions we suffered nothing more than a great jolt. We then bounced up a little and I thought we should decide to come down at

Caught in the thunderstorm

CAUGHT IN A THUNDERSTORM

once, but Willows said: 'Provided the guide-rope is touching the ground there is no real danger. I always go on until I get to a place where there is a train', and he pulled out an ABC timetable. We realised we were going in the direction of Hitchin. After one or two little ups and downs to carry the greater part of the guide-rope over cottages or telegraph wires, we finally came to a field close to a main road and saw that Hitchin was only about a mile away. Willows said: 'We will land here,' and coming down to about 20 feet from the ground he pulled the cord attached to the ripping-panel. At once a great slit appeared in the side of the balloon, allowing the gas to escape. A slight breeze carried it clear of the basket, it fell flat on the ground and we came down beside it. It was a most beautifully performed descent. We landed very smoothly and stepped out of the basket. Willows, who had the whole drill at his finger-tips, got us to help him stretch out the balloon and then neatly fold it up and fit it – or the greater part of it – into a large waterproof sheet; when it was put into the basket it bulged out at the top, looking rather like a very large loaf of bread. He then took from his pocket a large label addressed to the Spencer Balloon Company and

Landing after pulling the ripping panel

tied this to the basket. When that was done we walked up to the farmhouse, telling the farmer, who was a little surprised, that we had just landed in his field. We gave him some money to take the balloon and basket in one of his carts to Hitchin, the next time he was going there, and send it on by train to Spencers. Afterwards we set off for Hitchin. There was plenty of time before our train to London, so we celebrated our safe delivery from the thunderstorm with glasses of ale at a local pub. On arrival in London Willows asked me if I would mind talking to a friend of his on the staff of the *Daily Sketch*. He was at that time planning with the Spencer

Company to organise and to have built a very large captive balloon to carry people up to view London from one of the parks. He was full of imaginative ideas and also thought that on days of fog people would like to be taken up into the clearer sunshine above. I doubted whether he would get much profit by that. Anyway, when Willows got hold of his man on the telephone he asked me if I would give a description of our adventures in the thunderstorm, after which he would go on to explain about his balloon. I gave a brief outline, but, as you can imagine, the journalist got most of the facts wrong and the account was somewhat exaggerated. However, I reproduce the first part of his article as it appeared the following morning in the *Daily Sketch*. I do so because it explains the opening lines of an epic poem which Neil Mackintosh wrote of our adventure. He called it 'The Balloonatics', and this you will find at the end of the book, following the last chapter of our story.

CAUGHT IN A THUNDERSTORM

DAILY SKETCH Monday, October 6, 1924

BALLOONING THRILLS
LONDON PARTY'S ADVENTURES IN THUNDERSTORM
EXCITING WEEK-END

Enthusiast's Scheme to Lift City People Out of Fog

A perilous adventure yesterday befell a party of balloonists who left London on Saturday for a week-end in the air.

While among the clouds, 5,000 feet above the earth, they were overtaken by a thunderstorm and were forced to make an exciting descent.

Captain E.T. Willows, the well-known aeronautical expert, was the pilot, and there were two passengers in the balloon, which is of the Stream Line type.

Mr. A.C. Hardy, of Chelsea, one of the passengers, gave a graphical description of the adventure to the *Daily Sketch*.

'Our journey on Saturday brought us as far as Oxford, where we stayed the night. We started

back from Oxford at noon yesterday intending to return to London in the balloon.'

Mountains of Cloud

'We made a fine ascent, rising to a height of 10,000 ft. among great mountains of cloud which enveloped us and hid us from the earth. Then we dropped to about 6,000 ft., and were still among masses of cloud, many of which were dark and threatening. Rain began to fall, and we saw flashes of lightning followed by peals of thunder. Each time there was a burst of thunder the balloon trembled as though in response.

'The situation had become suddenly alarming in the extreme, for a thunderstorm is the balloonist's worst enemy. If the balloon had been struck by lightning we would have perished instantly. . . .'

CHAPTER SEVEN

Willows' Death in a Tragic Accident

CHAPTER SEVEN

This last chapter is a sad, indeed tragic, epilogue to the story. On August 3rd 1926 Willows was killed in a dreadful balloon accident while taking people up in a captive balloon at a Flower Show at Kempston Park near Bedford. I was then out on the 1925–27 *Discovery* oceanographic expedition to the Antarctic and it was only months later that I learnt of the disaster in newspapers reaching us at South Georgia.

A wind was getting up and Willows declared that this would be the last trip he would make. The balloon was anchored to a steel cable controlled by a power winch below. He was going up at the side of the Flower Show, not far from a group of trees, when quite suddenly the wind changed direction. He shouted to the man in charge of the winch to haul them down, but the balloon and cable drifted over to the trees and the cable stuck in a fork of a tree. They tried hard to free it, but in the meantime the wind rose

still further, and then the awful thing happened. The pressure of the wind on the balloon was too great for the netting which held it. The netting began to tear. Willows' last words, as recorded in the *Daily Mail*, were: 'For God's sake haul us down.' The next moment the net tore completely. The balloon shot up and the basket, with Willows and passengers, dropped like a stone.

I will now give the account which appeared on August 4th in *The Times* and then, also from *The Times* of the following day, the account of the Coroner's inquest.

THE TIMES Wednesday, August 4, 1926

BALLOON ACCIDENT AT FLOWER SHOW
FOUR KILLED IN FALL TO EARTH

Four persons were killed and one seriously injured this afternoon in an accident to a captive balloon which was making ascents at the annual flower show and sports meeting at Kempston, Bedford. The envelope of the balloon, while at a height of between 60 and 100 feet, broke away from the basket, which fell on to the field. Several thousand persons saw the accident.

The dead are Mr. W. Francis Harbage of Church-walk, Kempston, and Mrs. Harbage, his wife; Mrs. E. Crowsley of Bedford Row, Kempston; and Mr. E.T. Willows, the pilot of the balloon. The pilot and Mr. Harbage were killed immediately. Mrs. Harbage and Mrs. Crowsley died later in the County Hospital, Bedford. Mr. Ernest Crowsley received very serious injuries and late last night it was not expected that he would recover.

The balloon was the property of Messrs. T.G. Spencer & Sons of Highbury. The basket accommodated four passengers and the pilot, who occupied a hammock seat slung in the ropes suspending the basket to the envelope. The balloon was attached by a wire hawser to a winch attached to a motor lorry. Earlier in the afternoon it had made a number of successful ascents. Just before the accident, a little after 3 p.m., the balloon began an ascent with the pilot and four passengers. It rose to a height of 600 feet, and then the winch was started to wind in the rope. All went well until the balloon was about 60 to 100 feet above the ground, and then the hawser fouled a tree. At the same time the wind appeared to freshen and the balloon began to swing and tug at the rope.

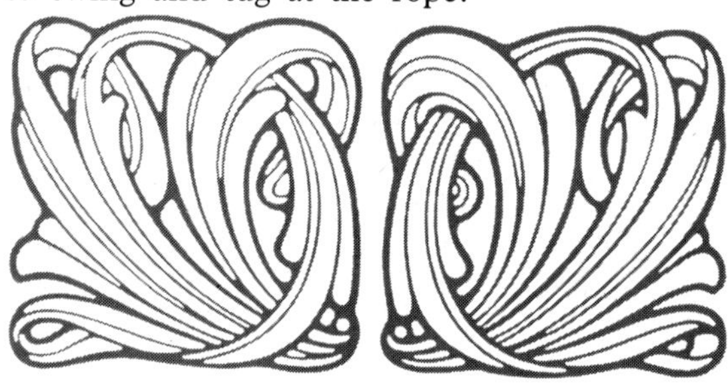

The winding gear did not seem strong enough to hold it down, and people on the ground took hold of the rope with the idea of helping to pull it in. The pilot could be seen in the ropes between the envelope and the basket and was heard shouting. His words, however, could not be distinguished. The man at the winch paid out more rope to ease the strain, and almost immediately the network of ropes between the basket and the envelope was seen to part. The envelope shot upwards, for the moment carrying the basket with it. Then the basket dropped and the pilot was seen either to jump or to fall clear of it. He fell on his head and his neck was broken. Mr. Harbage was also found to be dead.

A passenger who had made one of the earlier ascents said that the ropes seemed to be in good order. Describing the ascent, he said: 'I saw the balloon up in the air swinging about in what seemed to be a dangerous way. I could see the pilot in the rigging and I heard him blow a whistle. He then started shouting, and hundreds of people rushed forward and took hold of the ropes to try to pull the balloon down, because the winch did not seem strong enough to do so, as the wind was blowing pretty strongly. The balloon pulled and jerked and swung about in

the wind at a dangerous angle. I heard no cries from the people in the basket; only the pilot was shouting, and we could not hear what he said.

'The man at the winch decided to pay out some rope to ease the strain, and almost as he did so we saw the rope give way and heard a loud crack. The basket then fell like a stone. The pilot, I thought, jumped out, because he came down apart from the basket. Several people in the crowd fainted. We fetched motor-cars and got the injured away. The pilot was already dead and so was Mr. Harbage. Mrs. Harbage was still breathing and Mrs. Crowsley and Mr. Crowsley were alive too. The two bodies were taken to Kempston Mortuary and the other three were taken in motor-cars to the County Hospital in Bedford.'

An enquiry into the accident will be opened tomorrow afternoon.

THE TIMES Thursday, August 5, 1926

Evidence at Inquest

An inquest was held yesterday afternoon on the bodies of Mr. Ernest T. Willows and Mr. W.F. Harbage, at which the jury returned a verdict of 'Accidental Death.'

Mr. John Smith, a balloon rigger, of Highbury, employed by Messrs. Spencer & Sons, said that he was attending to the balloon at about 3.30 p.m. on the day of the accident. A gust of wind blew as the balloon was coming down. He noticed the trailing rope was caught in a tree, and six men were instructed to prevent the balloon from swaying. They could not move the balloon on account of the increasing wind, and he saw that something was happening to the net. He went across to assist in getting the balloon to the nearest point to the earth, and he was assisted by a crowd of up to one hundred men. The result of the increased strain on the trail rope cause it to break, but the winch rope held all the time. The net was tearing further and further, and the gas envelope was pressing through. His idea was that the net began to break where the parachute attachment was fixed.

Asked if there were no precautions taken to safeguard the passengers, the witness said that such an accident was almost unheard of. It was due to the heavy strain which came through a strong wind blowing at 30 miles an hour. The wind was 'full of rain', which made it 'press harder than any ordinary wind.' The witness added that the nets had been properly examined

on Saturday, Sunday and Monday. It was in its second season and was good. Its ordinary life would be four or five years.

Mr. Philip Mentor, another rigger, of Hackney, said that it would be fair to say that the assistance given by the crowd made the rent in the netting worse.

Mr. Ernest Allen, managing director of Spencer & Sons, said the net flew in the Gordon Bennett national race last year for the first time. It was the best obtainable. He had asked for the fullest possible enquiry. Mr. Willows was one of the finest pilots and he had taken up 1,000 passengers at Catford quite recently.

A policeman said that the balloon rope and the trail rope cleared the tree, enabling some men to get hold of the rope. They pulled it towards the winch, when suddenly a strong gust of wind came and blew it further over to the right. He noticed a big rent in the balloon, and a part of the net broke away. There was a noise like the flicking of sheets. The pilot shouted 'Pull it down!' and soon a hundred men were pulling. The pilot shouted again 'Pull it down: work hand over hand!' but suddenly the trail rope broke close to the balloon. The envelope went through the break and escaped. The basket and

the pilot's seat crashed to the ground. He cleared the people away, and found the pilot dead and Mr. Harbage dying.

It will be seen that at the inquest the representative of Spencer & Sons asked that a full enquiry be made into the cause of the accident. This was reported in *The Times* of Tuesday, November 2, 1926, and the following are extracts from that enquiry.

'At Bedford Police Court yesterday C.G. Spencer & Sons Limited, aeronautical engineers, of Highbury, London were summoned for allowing a balloon to be flown at Kempston, near Bedford, on August 3, without special permission in writing of the Secretary of State, and without it being certified as airworthy. . . .

'On the day of the accident there was no certificate of airworthiness, and no permit had been given. . . . [There was only a temporary certificate.]

'Major J.P.C. Cooper, Inspector of Accidents for the Air Ministry, said that the balloon had been registered as a spherical free balloon and

not as a captive balloon. The distinction was of great importance, for the stresses and strains imposed on the netting of the free balloon was very much less than upon a captive balloon. . . .

'The witness examined the cordage of the wrecked balloon on the day after the accident. He could not, however, say whether, had the cordage been examined before the accident, the balloon would have been passed as airworthy. . . .

'Answering further questions, Major Cooper said he found no evidence of mildew in the netting. There was fraying, which was due to the accident, and also fraying where the netting had not fractured. The evidence was fairly conclusive that the rope of the netting was made before 1919. . . .

'Mr. Cassels [K.C.]: Was the quality of the netting suitable for a captive balloon flight?
The witness: under the circumstances, yes. . . .

'. . . no one would suggest that failure to have a permit or a certificate had anything to do with

the accident, which was caused by an abnormal strain which no one could have foreseen. . . .

'. . . in the case of the failure to have a permit they imposed a fine of £100. . . .

With regard to the failure to have an airworthiness certficate . . . they imposed a penalty of £50.'

I would like to refer readers again to Mr. Alec McKinty's book *The Father of British Airships : a Biography of E.T. Willows*. There he gives an excellent account of the struggles of Willows to gain recognition and support for his genius in airship design; he was the first to produce a really efficient and navigable airship.

I would like to end my story by saying that apart from his being one of my schoolboy heroes, it was a great privilege to have met Willows, if only for such a short time. His infectious enthusiasm and his delightful sense of humour – one might say of fun – were aspects of his character that I shall always remember with gratitude for his sharing it with me. I hope this little book may have shown something of that character and his love of adventure.

APPENDIX

THE BALLOONATICS

An Epic Poem
by
Neil Mackintosh

Canto I

Attend all ye, who having read
What the 'Sketch' reporter said,
(An unimaginative bird
Who quoted Hardy word for word)
Would still aspire, in verses tense,
To learn the why and where and whence,
The whither, wherefore, how and when
A band of rash and reckless men
Sallied forth one afternoon
And beat it in a large balloon.

The day was fine and clear the sky
As to Vauxhall we did hie.
When scarce we'd filled the bulb with gas
The gasworks boss (the silly ass!)
Appeared and said to our surprise:
'When you let that darned thing rise
My best gasometer you'll hit
And that would cost you quite a bit.
I'd have you know,' pursued His Nibs,
'Gasometers have tender ribs.'
We answered: 'dammit, dash it, dammit,

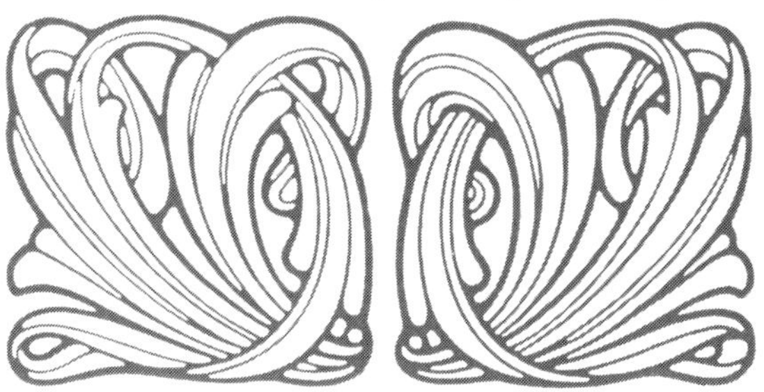

That will mean we'll have to cram it
On to that decrepit Ford,
And when thereon it's safely stored,
To drive across to Bugsby Bogs,
Somewhere near the Isle of Dogs.'
We reached the gasworks recommended,
Had the envelope distended,
Whilst the daylight faded fast,
So that when at long, long last
We were ready for our quest,
The sun was sinking in the west.

In the basket we were five,
And standing round some men did strive
To hold our chariot of the skies
Down to the ground till we should rise.
The pilot shouted: 'Let her go,'
And like an arrow from the bow
The great balloon, with airy grace,
Leaped gaily from the starting place.
'What Ho!' we cried, 'Behold the docks!
The ships, the shops, the traffic blocks,
The river's curve, the churches' spires,
St. Paul's, the Tower Bridge, Blackfrair's.'
And now we saw beneath our feet
The Abbey and Victoria Street,
And staring crowds below, we saw,

With rubber neck and gaping jaw.
But now, as darkness hid the sights,
The scene gave way to twinkling lights.

Now in the basket we had stored
Sufficient grub for all aboard,
But thoughtless of our drink and food,
We heeded not on what we stood,
Yet later, feeling somewhat void,
We were very much annoyed
To find adhering to our shoes,
A slimy pulp of yellow ooze.
Ripe bananas, crushed by feet,
Are not the best of things to eat.

The Kensington Museums we passed,
And then a little sand we cast,
Which fell like rather gritty snow
On to the streets and roofs below.
Unburdened thus, with boisterous bound,
Three thousand feet above the ground
The airship rose, while round about
The landscape swiftly opened out.
The hoot of cars, the bark of dogs,
The grunting of asthmatic hogs,
The ring of bells, the engines' blare,
Rose clearly on the quiet air.

Nearby, in brilliant lights arrayed,
The Exhibition was displayed.
And all the while the great balloon
Sailed on beneath a silvery moon.

Swiftly the hours sped away
As we pursued our lofty way.
Villages and towns we crossed,
Our bearings were completely lost.
So we approached the earth again
To see if we could ascertain
Our situation, more or less,
In this benighted wilderness.
We cried aloud to those below:
'Excuse me, Sir, we'd like to know
Our whereabouts. And we desire,
If we may venture to enquire,
To ascertain, if we came down,
How many miles we'd be from town.'
Some to answer did disdain,
Others with bibulous refrain,
Or unintelligible yell
Our courteous questions did repel.
Others again, with panic stirred,
When the celestial shouts they heard,
Bolted indoors and barred the door,
Whilst the clammy sweat did pour,

Over their faces, white and drawn,
Wishing they had ne'er been born.
But persevering in our query,
Never ceasing, never weary,
A man less dense in time we passed,
'High Wycombe' came the cry at last.

By this the hour was half-past eight,
And thinking it was rather late,
We cast our eyes on either hand
To see where we could safely land.
While thus we did deliberate
It chanced through some strange whim of fate
That we should not descend just now,
(The gods such plans would not allow)
But that the earth should rise on high
And meet us as we wandered by.
For sudden out the shades of night
A monstrous hillside loomed in sight;
The ground rose up and hit us – Crump!
The basket landed with a bump.
A moment later strange to say
We found to our intense dismay
That we were swinging to and fro,
The earth a thousand feet below.
But this was not the last we saw
Of those hills. There lay in store

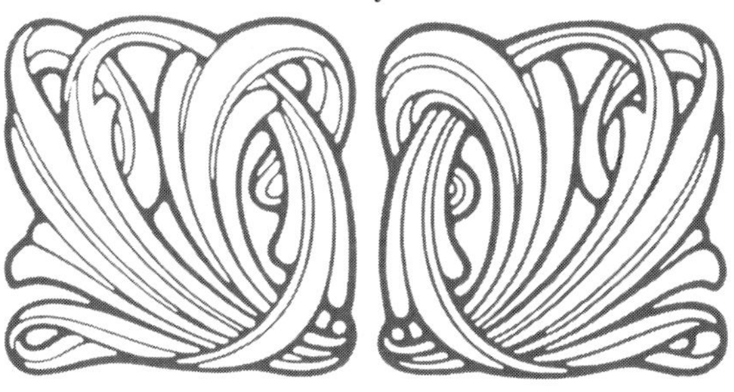

An unforeseen and deadly trap
Which might have caused a grave mishap.
'Grave' is a word appropriate,
For just before it was too late
We saw emerging from the gloom,
Encompassed by grave and vault and tomb,
A church upon a hill so high,
Its steeple seemed to reach the sky.
And as with fear we did perspire,
Lest we be spiked upon the spire,
The ballast bags we quickly manned;
The graves received a little sand
Instead of corpses, crushed and gory,
Else you'd not have heard my story.

Now heavenwards, rocket-like, we shot,
Left far below that grisly spot,
Three thousand feet we rose aloft,
Drifting through the moonlight soft.
The hills and trees had given place
To interplanetary space,
A death-like silence wrapped us round,
The ghostly void brought not a sound.

The time crept on till with a shock
We found that it was ten o'clock.
'Let us descend to earth,' we said,

'Our friends will think we're long since dead.'
And hovering soon o'er hedge and ditch,
Trying to make out which was which,
We spied a meadow, wide and green,
Bordering a winding stream.
A pleasant place it seemed to land,
A hamlet's lights were near at hand.
We quickly let escape some gas
And steeply fell towards the grass.
We landed with an awful splosh,
A beastly squelch, a horrid squash;
In haste we hurled out sandbags three,
And with a wrench we drifted free.
The clinging mud fell back in drips,
The quagmire seemed to lick its lips,
It quivered, gave a sucking sound,
And looked once more like firm dry ground.

We blundered through a grove of trees
And rose to meet the welcome breeze,
And when another mile was passed
We found a resting place at last.
For passing o'er a lonely road
Which the moonlight feebly showed,
Another field beyond, saw we;
No hill nor church nor marsh nor tree
Nor other obstacle there lay

To stop us landing straight away.
The grappling hook was quickly dropped,
The basket bumped and dragged and hopped,
And came to rest against a hedge
Bounding the field's further edge.
And here we tied her down with rope
To trunks and branches in the hope
That we should find her there next day
When ready to resume our way.
(The time was, if you'd like to know,
A quarter to eleven or so.)
These preparations being made,
The pilot in the basket stayed,
While we set forth to hunt around
And find where we had come to ground.
We met two yokels, and enquired
Our whereabouts. It then transpired
That Oxford was the nearest town,
Ten miles from where we had come down.
The nearest village bore the very
Curious name of Waterperry,
And Wheatley (rather larger) lay
A distance of three miles away.
We set off at a lively pace,
And soon we reached the last-named place,
And there we did contrive to rouse
The keeper of a public house,

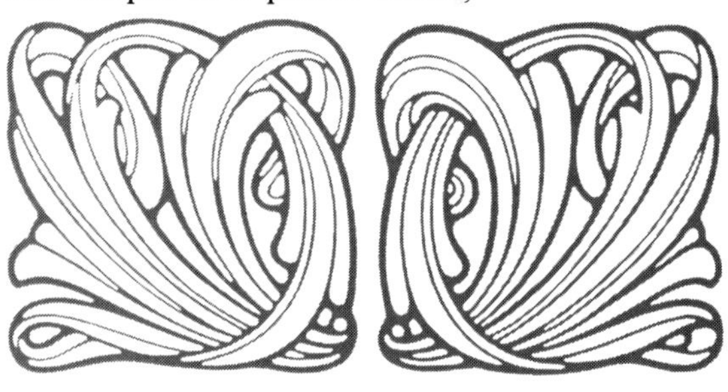

And as the distance was not far
We drove to Oxford in his car.
Two of us then took the train
In order to get home again
And two of us, without delay,
Returned to Wheatley, there to stay
And sleep until the sun should rise
To herald further enterprise.

Canto II

O ye, who on the Sabbath morn,
Arise at ten with lazy yorn,
Attempt to rouse yourselves in vain,
And then turn round and snore again,
Note the reward reserved by heaven
For us who breakfasted at seven.
Listen while the tale I tell
And hear what wondrous things befell.

The morning was distinctly fine,
The sun did condescend to shine,
White woolly clouds were piled high
Here and there about the sky,
The wind, which on the other day
Had blown us westward all the way,
Had now changed round as though it fain
Would blow us back to town again.
The news of our descent last night
Had filled the parish with delight,
And drew large numbers to the ground
Where in a paddock fenced around

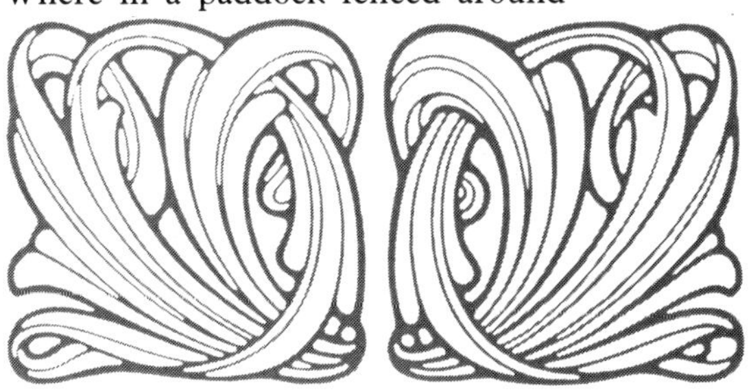

The famed balloon was situated,
Looking a trifle dissipated.

Once more we climbed aboard our craft
(The ship which lacked a fore and aft).
A few false starts she made at first,
Evincing a consuming thirst
To batter down a wretched tree
Which stood nearby quite harmlessly.
But finally midst plaudits loud
She rose above the watching crowd.
She cleared the hedge, she missed the trees,
And gaining courage by degrees,
Slowly at first but gath'ring pace,
She mounted into empty space.
Higher and higher did she go,
Until in half-an-hour or so,
Six thousand feet of country air
Divided us from Oxfordshire.

Now underneath the layer of food
And other things on which we stood,
There lay a monstrous coil of rope
For which we now began to grope,
And having tied the end up tight,
We hung it downwards that it might
Perform the function of a tail;

(This rope, when 'carried at the trail'
Gives special entertainment to
The navigator of 'the blue'.
For instance, it's a tempting bait. . . .
But let me not anticipate.)

Ascending heavenwards once again
We soon approached the cloud's demesne.
Past solid banks of mist we soared,
White hills and valleys we explored,
Until we sailed in triumph proud
Above the topmost layers of cloud.
Ten thousand feet our altitude,
And east the course which we pursued.
We could no longer see the ground,
The clouds were swiftly closing round.
They rose in twisted columns weird.
White, towering mountains had appeared;
Our upward lift was at an end
And now we started to descend.
Through mist three hundred fathoms deep
We passed as earthwards we did creep.
Our speed, though why I cannot tell,
Increased the further down we fell.
Descent became a headlong drop.
We threw out sand but could not stop.
The earth had reappeared at last;

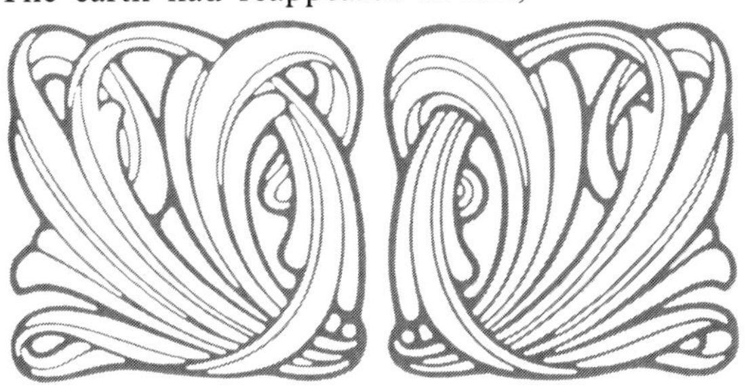

Another layer of cloud was passed.
And now in desperate haste we tried
To save ourselves from suicide.
We seized on every blessed thing –
Banana skins and bits of string,
Old bottles, paper bags and all –
We hurled them out to ease our fall.
T'was all in vain, for with a dash,
The earth rushed up. There came a crash,
A jolt which shook us through and through,
And nearly snapped our legs in two.
Again we rose with quick rebound,
But trailing now upon the ground
The rope, referred to herebefore,
In harmony with Newton's law
Inhibited our swift ascent
The more its weight on us was spent,
Till equilibrium was found
A gross of feet above the ground.
Thus wafted by the gentle breeze
We drifted over fields and trees,
Past woods and commons we did trail,
We wandered over hill and dale.
'Ere long there hove in sight below
A group of men, a score or so.
They saw the trail rope hanging out
And thought: 'There cannot be a doubt

That that balloon is going to land.
It's up to us to lend a hand.'
The rope, which they essayed to chase
Was setting quite a lively pace.
The first man seized the rope's end which
Deposited him in a ditch.
The next to catch it tried to halt
And turned a nasty somersault,
And as the rest with one accord
Pounced forward, we let fall toward
The ground a pint or so of sand
Just before they could lay hand
Upon the rope, which with a bound
Rose twenty feet above the ground.

Once clear we rose with leisure slow
Towards the clouds, now drifting low.
We little thought what peril lay
Behind, around, and in our way.
Meanwhile, like monstrous mountains white,
Their bases black as Stygian night,
The clouds banked up with silent gloom,
Like omens of impending doom.
The earth lay far away below,
The wind had almost ceased to blow,
When all at once a distant sound,
Like moving trains beneath the ground,

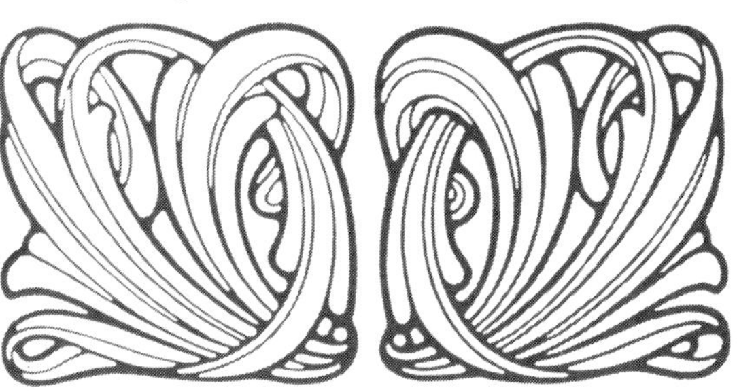

Did reach our horror stricken ears,
And filled our thoughts with sudden fears,
With visions of the lightning's flash,
Annihilated, fiery crash,
Heads, limbs and bodies strewn around,
And smouldering wreckage on the ground.
Again the grumbling thunder spoke
And through the clouds the lightning broke.
The valve above we opened wide,
The daft balloon at once replied
By going up instead of down,
Whilst all about, with angry frown,
The threatening clouds were closing round,
With some between us and the ground.
And though we pulled the valve again
Our efforts still were all in vain.
With lightning flashing, thunder pealing,
Inspired with that Kruschen feeling,
The frail balloon increased its height
Till all the world was out of sight,
Then started earthwards once again
Enveloped in the storm and rain.
It seemed that many an hour did pass
Before the trail-rope touched the grass,
But finally we reached the ground
Without mishap and safe and sound.

And now our flight was nearly done,
For it was but a few miles' run
To Hitchin, at which country town
We should for the last time come down.
We drifted on, with trail-rope out
For half-an-hour or thereabout,
Until the houses have in sight
To starboard of our line of flight.
While still a mile or so away
We suffered here a slight delay.
The wind had for the moment dropped,
And for a little while we stopped,
Hanging idly o'er the grass
Near to where a road did pass.
Now scanning the horizon, we
For some time not a soul could see,
But presently from every side
Where'er we looked a crowd we spied,
Converging on the field where
We seemed quite rooted to the air.
Just as again we felt the breeze,
The rope a hundred hands did seize.
With shouts of joy and screams of glee
They hauled us down that they might see
What kind of madmen can be found
In gas balloons, when brought to ground.
Now we did not propose to land

Until the town was near at hand.
Our pilot, an experienced bloke,
Inflated both his lungs and spoke.
A few well-chosen words he used,
Their ill-timed efforts he abused
In phrases plain and rather crude,
With epithets distinctly rude,
And recommended them to go
To regions tropical below.
The crowd, thus taken by surprise,
Released the rope and let us rise.
We drifted on a mile or so,
The rope still out and flying low,
Until we saw a main road wide,
Where, just upon the other side,
There lay a field, whose wide expanse
Attracted our enquiring glance.
We crossed the road and came straight down,
Quite near the outskirts of the town.
The trail-rope nearly wrecked a car
And gave a very nasty jar
To telegraphic wires which lay
Above the hedge of that highway.
Now hanging from the envelope
There dangled yet another rope.
The pilot, with a sudden spring,
Began to haul upon the thing.

We heard an awful rending sound,
The bulb subsided on the ground,
And with some last expiring sighs,
She crumpled up, no more to rise.
We packed her with consummate art
And stowed her on a passing cart.
A neighbouring inn we called at first,
And having somewhat quenched our thirst,
We staggered forth and took a train
Which brought us back to town again.